INTRUDER ALERT!

Based on the major motion picture
Adapted by Nick Rudd
Based on the Screenplay Written by Matt Lopez and Mark Bomback and Andy Fickman
Based on Characters Created by Alexander Key
Executive Producers Mario Iscovich, Ann Marie Sanderlin
Produced by Andrew Gunn
Directed by Andy Fickman

Printed in the United States of America
First Edition
1 3 5 7 9 10 8 6 4 2
Library of Congress Catalog Card Number on file.
ISBN 978-1-4231-1988-3
Visit disneybooks.com

Disney PRESS
New York

In the middle of the Nevada desert, a bright light filled the sky. Something was zooming toward Earth.

CRASH!

A spaceship slammed into the ground. On board were two aliens. Their names were Seth and Sara. They looked like normal teenagers.

MS267-GT-XYA-016   CONFIDENTIAL

PROJECT MOON DUST

Seth and Sara's arrival did not go unnoticed.
"We're registering impact!" an agent inside military
headquarters cried. He was looking at a bunch of
computer monitors.

"Get me Henry Burke on the phone," General
Lawson ordered. This was serious news!

In nearby Las Vegas, a cabdriver by the name of Jack Bruno was just trying to get through the day without any trouble. But that was hard to do. There was a UFO convention in town.

Back in the desert, Henry Burke was investigating the crash. He ordered his team to search the site for aliens. But Seth and Sara were no longer there. They had gotten on a bus headed for Vegas!

Jack Bruno was about to have a close encounter. One minute he was alone in his cab, and the next, there were two teenagers in his backseat! It was Seth and Sara. They needed his help.

"Where to?" Jack asked, turning on the meter.

Jack was not sure where the two kids had come from. They looked normal, but they had a funny way of talking. But he didn't mind. They were paying for the ride, after all.

Seth and Sara gave him directions, and Jack began to drive. But they were being followed! Henry Burke's men pulled up in black SUVs and tried to drive Jack off the road.

"Get down!" Jack cried. "Both of you!" They got away with some tricky driving and with the help of Seth and Sara's alien powers. But they weren't out of danger yet.

Jack drove for a long time. Finally, they arrived at a cabin in the middle of the desert. Seth and Sara paid for their ride. Then they sneaked inside the cabin. Jack decided to follow them.

The kids were hiding behind a couch. They ignored Jack and opened a refrigerator door. Behind it was a hidden staircase. Jack followed Seth and Sara down the stairs into a huge underground garden.

"What exactly is this place?" Jack asked.

Seth and Sara did not tell him the truth. They hid the fact that they were aliens. But they did tell him that they were there to get a special device. Others were after it, too!

Then they heard a thud.

Someone else was in the garden! Was it Burke?

It was worse! It was an alien creature called the Siphon! He had been looking for Seth and Sara. And now he was very, very angry.

Jack wasn't going to let the kids get hurt. He tackled the Siphon. Seth and Sara retrieved the device. Then, the three of them raced toward the stairs. From close behind, the Siphon shot laser beams at them.

Jack, Seth, and Sara made it out of the garden just before it blew up. It looked like the Siphon hadn't escaped. They were safe!

"You must not take such risks," Sara told Jack. She didn't want him to get hurt. Jack was just happy Seth and Sara were all right.

Jack was going to help them. He had no idea what that creature was, and he didn't know who those men chasing Seth and Sara were. But he was going to find out. And he had a feeling it was going to be a wild ride.